彗星、流星和小行星

太阳系的旅行者

U0240964

Comets, Meteors, and Asteroids

Voyagers of the Solar System

（英国）埃伦·劳伦斯／著　　刘 颖／译

江苏凤凰美术出版社

著作权合同登记图字：10-2022-144

图书在版编目（CIP）数据

彗星、流星和小行星：太阳系的旅行者 /（英）埃
伦·劳伦斯著；刘颖译 . -- 南京：江苏凤凰美术出版
社，2025. 1. --（环游太空）. -- ISBN 978-7-5741
-2027-3

Ⅰ. P185-49

中国国家版本馆 CIP 数据核字第 2024B5V297 号

策　　　划　朱　婧
责 任 编 辑　高　静　奚　鑫
责 任 校 对　王　璇
责任设计编辑　樊旭颖
责 任 监 印　生　嫄
英 文 朗 读　C.A.Scully
项 目 协 助　邵楚楚　乔一文雯

丛　书　名　环游太空
书　　　名　彗星、流星和小行星：太阳系的旅行者
著　　　者　（英国）埃伦·劳伦斯
译　　　者　刘　颖
出 版 发 行　江苏凤凰美术出版社（南京市湖南路1号 邮编：210009）
印　　　刷　南京新世纪联盟印务有限公司
开　　　本　710 mm×1000 mm　1/16
总 印 张　18
版　　　次　2025 年 1 月第 1 版
印　　　次　2025 年 1 月第 1 次印刷
标 准 书 号　ISBN 978-7-5741-2027-3
总 定 价　198.00 元（全 12 册）

版权所有　侵权必究
营销部电话：025-68155675　营销部地址：南京市湖南路1号
江苏凤凰美术出版社图书凡印装错误可向承印厂调换

目 录  Contents

书中加粗的词语见词汇表解释。

Words shown in **bold** in the text are explained in the glossary.

天空里的光芒
A Bright Light in the Sky

想象一下，你正在夜晚散步。

Imagine that you're out walking at night.

你抬头望向天空，突然发现一道明亮的光线划过黑夜。

You look up into the sky and suddenly spot something bright zooming across the darkness.

似乎是颗恒星，但它移动速度很快。

It looks like a **star**, but it's moving fast.

那道光芒转瞬即逝。

Then the flash of bright light disappears.

你刚刚看见的是一颗划过夜空的流星！

You have just seen a **meteor** streaking across the night sky!

流星 A meteor

之所以被称为"流星"，是因为它们看起来像在太空中飞驰而过的星星。不过，流星并非真正的恒星。

People sometimes call meteors shooting stars because they look like stars zooming across space. Meteors aren't really stars, though.

太空岩石
Space Rocks

你看到的流星最初是个太空岩石天体，叫作"流星体"。

The meteor you saw in the sky started out as a rocky space object called a **meteoroid**.

流星体是从彗星或一种被称为"小行星"的大型太空岩石上脱落的小碎片。

Meteoroids are small pieces that have broken away from comets or large space rocks called **asteroids**.

大多数流星体都很小，像沙粒，但有些能大到1米宽。

Most are tiny, like grains of sand, but some can be as large as 1 m wide.

当流星体靠近地球时，它会飞入地球周围的气体中。

When a meteoroid gets close to Earth, it flies into the gases surrounding our planet.

这层气体被称为"大气层"。

This layer of gases is called the **atmosphere**.

一旦进入地球大气层，流星体就会燃烧并产生明亮的光。

Once inside Earth's atmosphere, the meteoroid burns up and creates a bright streak of light.

每天有数百万颗流星体飞入地球大气层并燃烧殆尽。有些流星体由岩石组成，其他的则是岩石和金属的混合物。

Millions of meteoroids fly into Earth's atmosphere and burn up every day. Some are made of rock, while others are a mixture of rock and metal.

大气层 **Atmosphere**

流星 **A meteor**

地球 **Earth**

这张照片展示了地球大气层中的一颗流星。该照片是由一名宇航员在国际空间站拍摄的。

This photo shows a meteor in Earth's atmosphere. The photo was taken by an astronaut on the International Space Station.

太阳系 The Solar System

在太空中，数十亿颗流星体环绕太阳运行。

Out in space, there are billions of meteoroids circling the Sun.

这些小岩石块与结满冰的彗星和小行星相伴。

These small rocky chunks are joined by icy **comets** and asteroids.

还有八大行星环绕着太阳。

There are also eight **planets** circling the Sun.

它们分别是水星、金星、我们的母星地球、火星、木星、土星、天王星和海王星。

The planets are called Mercury, Venus, our home planet Earth, Mars, Jupiter, Saturn, Uranus, and Neptune.

太阳及其周围的天体系统被称为"太阳系"。

The Sun and its family of space objects are called the **solar system**.

太阳系中的大多数小行星都集中在被称为"小行星带"的环状带中。

Most of the asteroids in the solar system are in a ring called the asteroid belt.

太阳系 The Solar System

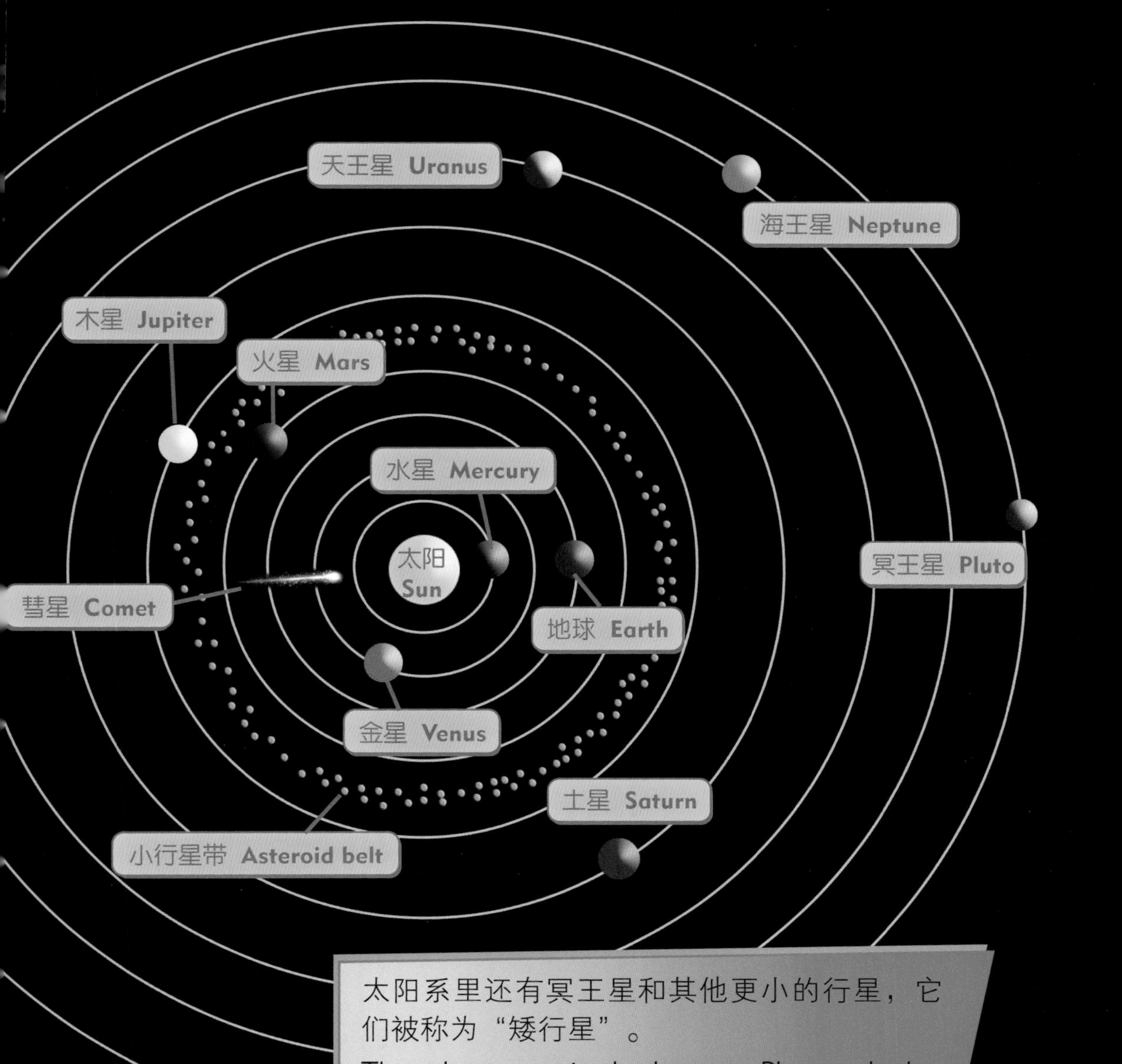

天王星 Uranus

海王星 Neptune

木星 Jupiter

火星 Mars

水星 Mercury

太阳 Sun

冥王星 Pluto

彗星 Comet

地球 Earth

金星 Venus

土星 Saturn

小行星带 Asteroid belt

太阳系里还有冥王星和其他更小的行星，它们被称为"矮行星"。

The solar system is also home to Pluto and other small planets, called **dwarf planets**.

小行星
Asteroids

大多数小行星的形状像土豆块。

它们有些小得像辆汽车，有些大得像座山。

在小行星带中，数以百万计的小行星围绕太阳公转。

这些巨大的太空岩石时不时就撞在一起。

撞飞出来的小石粒，就成了在太空里穿梭的流星体。

Most asteroids are shaped like lumpy potatoes.

They can be as small as a car or as large as a mountain.

There are millions of asteroids circling the Sun in the asteroid belt.

Sometimes these large space rocks crash into each other.

Smaller pieces break off and become meteoroids traveling through space.

小行星由岩石和金属混合而成。这颗小行星名叫司琴星。

Asteroids are made of rock mixed with metals. This asteroid is called Lutetia (loo-TEE-shee-yuh).

流星体 Meteoroid

小行星 Asteroid

这幅图展示了小行星被撞碎的情景。这些小碎片就是流星体。

This picture shows how an asteroid might look as it breaks into pieces. The small pieces are called meteoroids.

巨型小行星
Asteroid Giants

小行星带中最大的小行星是谷神星。

The largest asteroid in the asteroid belt is named Ceres (SIHR-eez).

它宽达952千米，几乎和美国得克萨斯州一样大！

At 952 kilometers across, it is almost as wide as Texas!

灶神星也是一颗巨大的小行星，在小行星带中围绕太阳公转。

Vesta is also a gigantic asteroid that **orbits** the Sun in the asteroid belt.

很久以前，灶神星撞上了另一个大型天体。

At some time in the past, Vesta crashed into another large space object.

尘埃、碎石和很多山体大小的小行星碎片被抛入太空中。

Dust, rocks, and mountain-sized chunks of the asteroid were thrown into space.

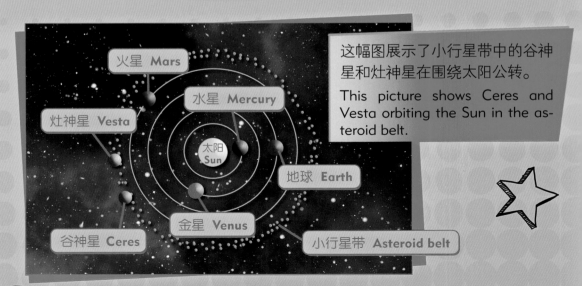

这幅图展示了小行星带中的谷神星和灶神星在围绕太阳公转。

This picture shows Ceres and Vesta orbiting the Sun in the asteroid belt.

这幅图展示了谷神星与地球的大小对比。谷神星既是小行星，也是矮行星。

This picture shows the size of Ceres compared to Earth. Ceres is an asteroid and a dwarf planet.

谷神星 Ceres

地球 Earth

灶神星 Vesta

这是灶神星的照片。它宽达530千米。科学家认为，灶神星的一些碎片曾穿越太空并坠落在地球上。

This is a photo of Vesta. It is just over 530 kilometers wide. Scientists think that some pieces of Vesta have traveled through space and landed on Earth.

陨石
Meteorites

有时，进入地球大气层的天体并未燃烧殆尽。

Sometimes, space objects that enter Earth's atmosphere do not completely burn up.

有些可能从这火热的旅程中幸存下来并坠落在地球上。

They may survive their fiery voyage and land on Earth.

降落在地球上的天体被称为"陨石"。

A space object that lands on Earth is called a **meteorite**.

陨石可能由岩石或金属组成，或由二者混合而成。

Meteorites may be chunks of rock or metal, or a mixture of both.

有些陨石只有鹅卵石大小，而有些则有篮球大小甚至更大。

Some are just the size of a pebble, while others are basketball-sized or even larger.

科学家发现了数枚陨石，他们认为这些陨石是灶神星的碎片。

Scientists have found several meteorites that they believe are pieces of the asteroid Vesta.

科学家认为这枚陨石来自灶神星。它宽约10厘米。

Scientists think this meteorite came from Vesta. It is about 10 centimeters wide.

地球上已发现的最大陨石位于非洲。它就是霍巴陨石。它的主要成分是铁，整体和一辆汽车一样大！

The largest meteorite that's been found on Earth landed in Africa. It's called the Hoba meteorite. It is mostly made of iron and is as big as a car!

霍巴陨石 **Hoba meteorite**

这枚陨石和地球上的石头相似。

This meteorite looks similar to a stone from here on Earth.

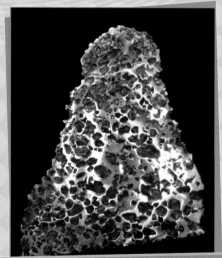

可以很明显地看见这枚陨石中的金属。

It's easy to see the metal in this meteorite.

来自太空的危险
Dangers from Space

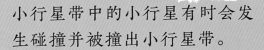

小行星带中的小行星有时会发生碰撞并被撞出小行星带。

Sometimes asteroids in the asteroid belt collide and get knocked outside of the belt.

它们可能因此在距离地球更近的轨道上绕太阳公转。

Then they may begin circling the Sun in a pathway much closer to Earth.

很久以前，一些巨大的小行星曾与地球相撞。

In the past, large asteroids have collided with Earth.

科学家密切关注在地球附近公转的所有大型天体。

Scientists keep watch on any large objects that are orbiting near our planet.

他们能够追踪这些天体运行的轨道和将要去的地方。

They can track the pathways these objects are taking and where they will go.

科学家确信，目前已被发现的小行星都不会与地球相撞。

Today, scientists are sure that none of the asteroids they've discovered will collide with Earth.

陨石坑 Crater

美国亚利桑那州的这处陨石坑宽约1200米。大约5万年前，一个大型天体与地球相撞，留下了这处陨石坑。

This **crater** in Arizona is nearly 1,200 meters wide. It was made by a large space object that collided with Earth about 50,000 years ago.

科学家在墨西哥希克苏鲁伯镇附近发现了一个巨大的陨石坑。它是由一个宽约10千米的天体造成的。这个天体在6600万年前撞击了地球。

Scientists found a giant crater near the town of Chicxulub (CHEEK-shuh-loob) in Mexico. It was made by a space object that was about 10 km wide. This object hit Earth 66 million years ago.

这个天体撞上墨西哥后，产生了巨大且炽热的灰尘云。撞击还引发了地震，导致海水淹没了陆地。许多科学家认为此次撞击导致了恐龙和许多其他动物的灭绝。

The space object that hit Mexico sent up huge clouds of hot ashes. It also caused earthquakes and made oceans flood the land. Many scientists believe this event helped kill off the dinosaurs and many other animals.

巨大的太空"雪球"
Giant Space Snowballs

彗星是由冰、岩石和尘埃混合而成的大型球体。

Comets are large balls of ice mixed with rock and dust.

大多数彗星绕太阳飞行的轨道都很长，像鸡蛋的形状。

Most comets travel in a long, egg-shaped journey around the Sun.

当彗星远离太阳时，它是由冰、尘埃和岩石组成的"冰块"。

When a comet is far from the Sun, it is a cold chunk of ice, dust, and rock.

当它靠近太阳时，温度迅速升高。

As it gets closer to the Sun, it heats up.

接着，它会释放出一大团气体和尘埃。

Then it gives off a huge cloud of **gases** and dust.

气体和尘埃形成一条尾巴，这条尾巴能向外拖曳出数百万千米。

The gases and dust form tails that stretch behind the comet for millions of kilometers.

远离太阳时的彗星
Comet when far from Sun

靠近太阳时的彗星
Comet when near to Sun

这张图显示了彗星绕太阳公转的轨道。
This diagram shows a comet's orbit around the Sun.

太阳 Sun

这张照片显示了霍姆斯彗星周围的气体和尘埃云。这团尘埃云甚至比太阳还大！

This photo shows the gas and dust cloud around a comet called Comet Holmes. The cloud grew to be bigger than the Sun!

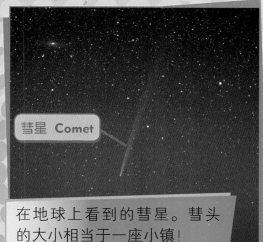

在地球上看到的彗星。彗头的大小相当于一座小镇！

A comet seen from Earth. The head of a comet can be the size of a small town!

通过望远镜看到的彗星。

A comet seen through a telescope.

奇幻的太空任务
Amazing Space Missions

2004年，一个名叫"罗塞塔号"的太空探测器被发射升空，它的观测目标是一颗叫作67P的彗星。

在10年的漫长旅程后，"罗塞塔"在2014年抵达了67P彗星并开始围绕它公转。

"罗塞塔号"放出了一台探测器，名叫"菲莱号"。它在彗星的冰冻表层上着了陆。

这是有史以来第一次有人造飞行器降落在彗星上！

"罗塞塔号"和"菲莱号"将搜集到的信息送回地球，让科学家得知了更多有关彗星的信息。

In 2004, a space **probe** named *Rosetta* began its mission to a comet called 67P.

After a 10-year journey, *Rosetta* reached 67P in 2014 and began orbiting the comet.

Rosetta released a robotic lander, called *Philae*, which landed on the comet's icy surface.

It was the first time that a human-made craft had ever landed on a comet!

Rosetta and *Philae* sent information back to Earth that will tell scientists more about comets.

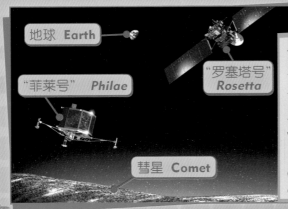

任务结束后，"罗塞塔号"坠毁在67P彗星上。现在，"罗塞塔号"和"菲莱号"都乘坐着这颗冰冷的彗星在太空中旅行。

When the mission ended, *Rosetta* crashed into 67P. Now *Rosetta* and *Philae* are flying through space aboard the icy comet.

这张67P彗星的照片是"罗塞塔号"拍摄的。

This photo of 67P was taken by *Rosetta*.

2019年，日本太空探测器"隼鸟2号"在一颗叫"龙宫"的小行星上登陆。它从这颗小行星的表面采集了岩石样本。

In 2019, a Japanese spacecraft named *Hayabusa 2* landed on an asteroid called Ryugu. The robotic spacecraft collected rocky samples from the asteroid's surface.

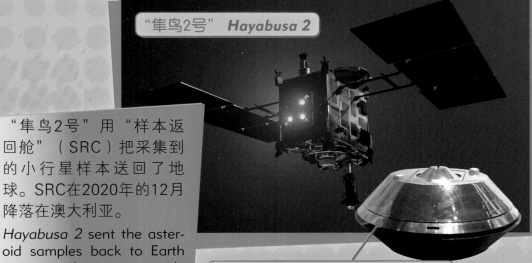

"隼鸟2号" *Hayabusa 2*

"隼鸟2号"用"样本返回舱"（SRC）把采集到的小行星样本送回了地球。SRC在2020年的12月降落在澳大利亚。

Hayabusa 2 sent the asteroid samples back to Earth in a sample-return capsule (SRC). The SRC landed in Australia in December 2020.

样本返回舱的模型
A model of a sample-return capsule

动动手吧：彗星和小行星游戏
Get Crafty : Comets and Asteroids Game

动手制作属于自己的彗星、小行星和流星体棋盘游戏。

在一大张正方形的绘画纸上画出25个方格。在方格里画出彗星、小行星和流星体。快制订游戏规则，并与朋友们一起玩吧！

游戏思路

你可以参考以下3点：

· 掷骰子并根据骰子的点数移动相应步数。当你落在某个方格时，看看那里有什么。
· 如果降落在彗星上，你可以向前移动3个方格。
· 如果降落在小行星上，则必须回到起点。

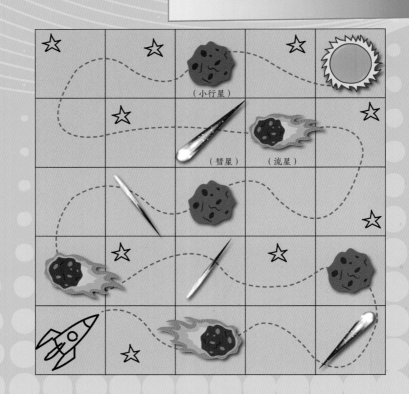

（小行星）
（彗星） （流星）

词汇表 Glossary

小行星 | asteroid
围绕太阳公转的大块岩石，有些小得像辆汽车，有些大得像座山。

大气层 | atmosphere
行星、卫星或恒星周围的一层气体。

彗星 | comet
由冰、岩石和尘埃组成的天体，围绕太阳公转。

陨石坑 | crater
圆形坑洞，通常由小行星和其他大型岩石天体撞击行星或卫星表面而形成。

矮行星 | dwarf planet
围绕太阳运行的圆形或近圆形天体，比八大行星小得多。

气体 | gas
无固定形状或大小的物质，如氧气或氦气。

流星 | meteor
小型天体（如流星体）进入地球大气时燃烧并产生明亮的光迹。

陨石 | meteorite

坠落在行星或卫星上的流星体残骸。

流星体 | meteoroid

从彗星或小行星上脱落的一小块岩石或尘埃。

公转 | orbit

围绕另一个天体运行。

行星 | planet

围绕太阳公转的大型天体：一些行星，如地球，主要是由岩石组成的；其他的行星，如木星，主要是由气体和液体组成的。

探测器 | probe

不载人太空飞船。通常被送往行星或其他天体，用于拍摄照片并收集信息，由地球上的科学家操作控制。

太阳系 | solar system

太阳和围绕太阳公转的所有天体，如行星及其卫星、小行星和彗星。

恒星 | star

燃烧的巨型气态星球。我们的太阳就是一颗恒星。